by Kathleen Powell

Copyright © by Harcourt, Inc.

All rights reserved. No part of this publication may be reproduced or transmitted in any form or by any means, electronic or mechanical, including photocopy, recording, or any information storage and retrieval system, without permission in writing from the publisher.

Requests for permission to make copies of any part of the work should be addressed to School Permissions and Copyrights, Harcourt, Inc., 6277 Sea Harbor Drive, Orlando, Florida 32887-6777. Fax: 407-345-2418.

HARCOURT and the Harcourt Logo are trademarks of Harcourt, Inc., registered in the United States of America and/or other jurisdictions.

Printed in Mexico

ISBN 978-0-15-362412-4

ISBN 0-15-362412-4

7 8 9 10   0908   16 15 14 13 12 11
4500336780

Visit *The Learning Site!*
www.harcourtschool.com

Introduction

How would you answer this riddle?

Sometimes I am a liquid.

Sometimes I am a solid.

Sometimes I am a gas.

You use me every day.

What am I?

If you answered water, you are right! Water can change to all of these forms.

But do you know how?

Water in three forms

Heat and Cold

Heating and cooling can make water change forms. The temperature of the water affects how the water changes.

Heating water to 100 degrees Celsius makes it boil. Cooling water to 0 degrees Celsius makes it freeze. You can find these temperatures on a Celsius thermometer.

Another temperature scale is called the Fahrenheit scale. On the Fahrenheit scale, water boils at 212 degrees. It freezes at 32 degrees. Weather reports in the United States usually describe temperatures using the Fahrenheit scale.

Evaporation

Here's another riddle.

What gets wetter the more it dries?

A towel! (As it dries you, it gets wet!)

What happens when you put a wet towel into a dryer? The heat from the dryer dries the towel. Remember, temperature affects water's form. What happens to the water that is in the towel? How does it change?

The heat makes the water evaporate, or change from a liquid to a gas. When water is a gas, it is called water vapor.

You can't see water vapor. It's invisible. Think of a kettle of water on a stove burner. When the water boils, you can see steam. Steam isn't made of just water vapor, however. Steam is also made of tiny drops of liquid water. Those drops are what you see when you see steam.

steam

What do you think would happen if the water in the kettle kept boiling? All of the liquid water would evaporate, or change to water vapor.

Condensation

You have seen how liquid water can change to water vapor. But can water vapor change back to a liquid? Let's find out.

Suppose that it's a hot day. You want to have a glass of cold water. You put ice cubes into a drinking glass and fill the glass with water. Then you let the glass sit in the warm air. What happens to the outside of the glass?

After awhile, you see that the outside of the glass has beads of water on it. Can the water be leaking through the glass?

What really happens is called condensation. Water vapor in the warm air turns into a liquid when it touches the cold glass.

condensation

More Condensation

Are you ready for another riddle?

What always falls but never gets hurt?

Rain!

You have seen some ways in which liquid water can change to a gas. What do you think happens to all of that water vapor in the air?

When the water vapor cools, it changes to drops of liquid water. The water drops form clouds. Just like the part of steam that you can see, clouds are made of tiny drops of water. If the drops get too big and heavy, the water falls to Earth as rain.

◀**Temperature Changes**
Depending on the temperature, water may fall as snow or rain.

Water Size

You have read about some ways in which water changes form. If water freezes, it changes to a solid. If water boils, it changes to a gas called water vapor. If the water vapor cools, it changes to a liquid. Did you know that the amount of space the water takes up also changes?

Suppose you put a container of liquid water in the freezer. Would the ice it changes to take up as much space as the liquid did? Most things take up less space in their solid form. But not water—it takes up more space, not less.

Solid Water ▶
Ice in your freezer takes up more space than liquid water.